YOUR KNOWLEDGE HAS VALUE

Bibliographic information published by the German National Library:

The German National Library lists this publication in the National Bibliography; detailed bibliographic data are available on the Internet at http://dnb.dnb.de .

Imprint:

Copyright © 2014 GRIN Verlag, Open Publishing GmbH
Print and binding: Books on Demand GmbH, Norderstedt Germany
ISBN: 9783668384262

This book at GRIN:

http://www.grin.com/en/e-book/351869/lattices-and-their-application-in-cryptography

Merve Cakir

Lattices and their application in Cryptography

GRIN Publishing

GRIN - Your knowledge has value

Since its foundation in 1998, GRIN has specialized in publishing academic texts by students, college teachers and other academics as e-book and printed book. The website www.grin.com is an ideal platform for presenting term papers, final papers, scientific essays, dissertations and specialist books.

Visit us on the internet:

http://www.grin.com/

http://www.facebook.com/grincom

http://www.twitter.com/grin_com

BACHELOR-THESIS

Lattices and their application in Cryptography

June 24, 2014

Institute of Computer Technology

Abstract

The aim of this thesis is to identify the characteristics of lattice-based cryptosystems. The use of encryption and signature schemes can be insecure considering attacks by a quantum computer and inefficient in the computation time. An alternative cryptography is proposed, which is based on worst-case lattice problems. The security and the hardness of the underlying computational problems will be analyzed by providing collaboration between the linear-algebra, complexity-theory and the public-key cryptography.

Contents

List of Figures

1 Introduction

The Internet has become an indispensable part in the work and private life. Thus, the necessity for secure communication continues to increase and became one of the most important tasks.

The public-key cryptography was published in 1976 and had awakened much interest. Cryptosystems could be developed based on mathematically hard problems and other applications such as digital signatures were worked out. The RSA is the most popular public-key encryption scheme, which provides convenience with respect to key exchange. It also gives the possibility of authentication.

However, a major risk of asymmetric cryptosystems is the security, which depends on computational problems such as factoring large numbers and discrete logarithm problem. Thus at a point, these systems are insecure.

In contrast, the use of lattices introduces problems, which are at least NP-complete and therefore also suitable for the cryptography.

This thesis provides an exciting opportunity to advance our knowledge of lattices and some lattice-problems, which are described by working out proofs step-by-step, to present the construction of new cryptosystems and analyze them. Examples are provided to make the material more comprehensible.

1.1 Outline

This thesis is categorized into two parts, which are as follows:

In the second chapter the fundamentals of the lattice-theory will be presented. We introduce lattice basis reduction with the corresponding LLL-algorithm and define lattice-problems that have been proved to be hard. We also provide a brief overview of public-key cryptography and digital signatures.

The third chapter describes the application of lattices in cryptography. We describe the construction of the lattice-based cryptosystem Goldreich-Goldwasser-Halevi and analyze the security in which we present the most effective attacks against it. We will conclude with a comparison to other cryptosystems like RSA.

2 Mathematical Background

In this chapter, the mathematical foundations will be explained, on which the lattice-based cryptography is constructed. In the first section, lattices and some lattice-based problems will be presented. We will also be concerned with the reduction of lattices. The second section provides a brief overview of public-key systems and digital signatures.

2.1 Lattices and Lattice Reduction Problems

The most definitions and theorems are based on the lecture notes of Claus-Peter Schnorr [1] and the lecture notes of Daniele Micciancio [2].

2.1.1 Definitions and Properties

Definition 2.1.1.1 (Lattice): *Let $b_1, ..., b_n \in \mathbb{R}^m$ be n linearly independent vectors. The set*

$$L(b_1, ..., b_n) = \left\{ \sum_{i=1}^{n} x_i b_i \mid x_i \in \mathbb{Z} \right\}$$

is called a lattice of rank n. The sequence of vectors $b_1, ..., b_n$ forms a basis of it and is denoted by the $m \times n$ matrix $B = [b_1, ..., b_n]$.

The set $L(B)$ defines a lattice which is spanned by the basis B. If $n = m$, the lattice is called a **full rank lattice**. Note that the definition of a lattice is similar to the definition of a vector space:

$$span(B) = \left\{ \sum_{i=1}^{n} x_i b_i \mid x_i \in \mathbb{R} \right\}$$

Compared to the vector space, where any arbitrary real coefficients can be combined, in a lattice only integer coefficients can be used.

Theorem 2.1.1.1: *A lattice L is a discrete additive subgroup of $\mathbb{R}^m$.*

Proof. Let $B = \{b_1, ..., b_n\}$ be a basis and $\varphi : \mathbb{R}^n \to span(B) \subset \mathbb{R}^m$ be a linear mapping with $\varphi(x_1, ..., x_n) = \sum_{i=1}^{n} x_i b_i$. Then $\varphi(\mathbb{Z}^n) = L$. The vectors $b_1, ..., b_n$ are linearly independent, thus the linear mapping φ is invertible, i.e. it is an isomorphism such that $\mathbb{Z}^n$ and L are isomorphic. Since $\mathbb{Z}^n$ is discrete, we have that L is also discrete.

$\square$

Theorem 2.1.1.2: *Each discrete additive subgroup of $\mathbb{R}^m$ containing n linearly independent vectors is a lattice.*

Proof. Let $L \subset \mathbb{R}^m$ be a discrete additive subgroup. Assume there exist n linearly independent vectors in L such that $n \leq m$. We show that L is a lattice of rank n over induction:

$n = 1$: assume $0 \neq b \in L$ is the shortest vector w.r.t. l_2-norm, then $L(B) = L$.

$n > 1$: choose $b_1 \in L \backslash \{0\}$ with $\frac{1}{k} b_1 \notin L \; \forall k \geq 2$. Then $L(b_1) = L \cap span(b_1)$. Using orthogonal projection $\pi : \mathbb{R}^m \to span(b_1)^{\perp}$ with $\pi(b) = b - \frac{\langle b, b_1 \rangle}{\langle b_1, b_1 \rangle} b_1$ we have to show that $\pi(L)$ is discrete and contains $n-1$ linearly independent vectors.

Recall that an additive subgroup is called discrete if 0 is not a limit point, so we have to show that 0 is not a limit point of $\pi(L)$. Assume the distinct vectors $\pi(y^{(i)})$ with $y^{(i)} \in L$ converge to 0. We obtain vectors $\bar{y}^{(i)} \in L$ with

$$\bar{y}^{(i)} = y^{(i)} - \left\lfloor \frac{< y^{(i)}, b_1 >}{< b_1, b_1 >} \right\rceil b_1.$$

We have $\bar{y}^{(i)} - \pi(y^{(i)}) = \left(\frac{<y^{(i)}, b_1>}{<b_1, b_1>} - \left\lfloor \frac{<y^{(i)}, b_1>}{<b_1, b_1>} \right\rceil \right) b_1$ and $\|\bar{y}^{(i)} - \pi(y^{(i)})\| \leq \frac{1}{2} \|b_1\|$. Because of $\lim_{i \to \infty} \|\pi(\bar{y}^{(i)})\| = 0$, there exist infinitely many vectors $\bar{y}^{(i)} \in L$ with $\|\pi(\bar{y}^{(i)})\| \leq \|b_1\|$, which is a contradiction to discreteness of L.

Thus, $\pi(L)$ is discrete.

To show that $\pi(L)$ is a lattice of rank $n - 1$, we show $L \subset L(b_1, ..., b_n)$:

For any lattice vector b we have $\pi(b) \in \pi(L) = L(\pi(b_2), ..., \pi(b_n))$. It follows that $\exists \tilde{b} \in L(b_2, ..., b_n)$ with $\pi(b) = \pi(\tilde{b})$ and it holds that $b - \tilde{b} \in span(b_1)$. Choose of b_1 yields to $L(b_1) = L \cap span(b_1)$ that means $b - \tilde{b} \in L(b_1)$. Thus for all bases $\{\pi(b_2), ..., \pi(b_n)\}$ of $\pi(L)$ it holds $L = L(b_1, ..., b_n)$.

$\square$

Definition 2.1.1.2 (Lattice-vector): *If a vector v belongs to L, it is called a lattice vector.*

After these definitions we can give a simple example of lattices in $\mathbb{R}^2$.

Example 1: *Let* $b_1 = \begin{pmatrix} 1 \\ 0 \end{pmatrix}$ *and* $b_2 = \begin{pmatrix} 0 \\ 1 \end{pmatrix}$, *which are linearly independent. The lattice spanned by them is* $\mathbb{Z}^2$ *(see Fig.2.1).*

Now consider the linearly independent vectors $v_1 = \begin{pmatrix} 1 \\ 0 \end{pmatrix}$ *and* $v_2 = \begin{pmatrix} 0 \\ 2 \end{pmatrix}$. *The lattice spanned by them is not* $\mathbb{Z}^2$ *since we can not generate e.g. the vector* $\begin{pmatrix} 0 \\ 1 \end{pmatrix}$ *(see Fig.2.2).*

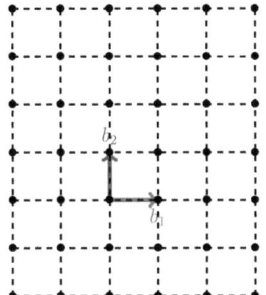

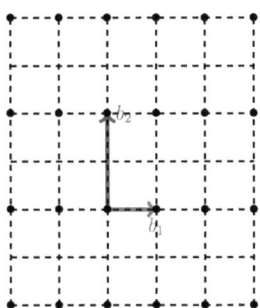

Figure 2.1: Lattice $\mathbb{Z}^2$ Figure 2.2: Basis can not generate $\mathbb{Z}^2$

This leads to the question when a set of linearly independent vectors can form a given lattice. To answer this question, we need the definition of the Parallelepiped.

Definition 2.1.1.3 (Fundamental Parallelepiped): *For any basis B, the parallelepiped is defined as*

$$P(B) = \{Bx \mid x \in \mathbb{R}^n, \forall i : 0 \le x_i < 1\}.$$

Theorem 2.1.1.3: *Let L be a lattice in* $\mathbb{R}^m$ *and* $B = \{b_1, ..., b_n\}$ *be a set of n linearly independent vectors. Then B forms a basis of L if and only if* $P(B) \cap L = \{0\}$.

Proof.

"$\Rightarrow$" Let B be a basis of a lattice L. By Definition 2.1.1.1, L is the set of all linear integer combinations of B and by Definition 2.1.1.2, $P(B)$ is the set of linear combinations of B with variables in $[0, 1)$. The only integer combination that belongs in $P(B)$ is the one where $x_i = 0$ for all $1 \leq i < n$. Thus $P(B) \cap L = \{0\}$.

"$\Leftarrow$" Assume $P(B) \cap L = \{0\}$ and B is not a basis. Then there exist a lattice vector v such that $v = \sum\limits_{i=1}^{n} y_i b_i$ for at least one not integral y_i. Consider the vector $v' = \sum\limits_{i=1}^{n} \lfloor y_i \rfloor b_i$. Since $\lfloor y_i \rfloor \in \mathbb{Z}^n$ the vector v' is also in L. By Theorem 2.1.1.1, L is an additive subgroup thus $v - v' = \sum\limits_{i=1}^{n}(y_i - \lfloor y_i \rfloor)b_i = x \in L$. With $(y_i - \lfloor y_i \rfloor) \in [0, 1)$ the lattice-vector x is in $P(B)$. Due to the assumption x must be the zero-vector. The linear independency of the vectors $b_1, ..., b_n$ gives that it must be $y_i - \lfloor y_i \rfloor = 0$. Since there exist at least one $y_i \in \mathbb{R}$ we have $y_i - \lfloor y_i \rfloor > 0$. Thus $x = 0$ if and only if B is a basis of L.

$\square$

In our previous example where $v_1 = (1, 0)^\top$ and $v_2 = (0, 2)^\top$ we have that $P(B) \cap L = \{(0, 0), (0, 1)\}$ so the set of v_1, v_2 does not form a basis of the lattice $\mathbb{Z}^2$. Now consider the linearly independent vectors $b_1 = (1, 1)^\top$ and $b_2 = (2, 1)^\top$. These vectors can also produce the lattice $\mathbb{Z}^2$ (see Fig. 2.3).

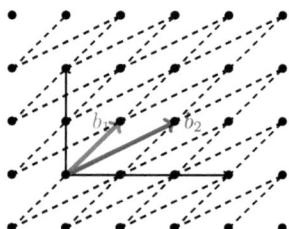

Figure 2.3: Lattice $\mathbb{Z}^2$ with another basis

To decide in general when two bases can generate the same lattice, further definitions are required.

Definition 2.1.1.4 (Unimodular): *A matrix $U \in \mathbb{Z}^{n \times n}$ is called unimodular if $det(U) = \pm 1$. Denote unimodular matrices by $GL_n(\mathbb{Z})$.*

Note that if the matrix $U \in \mathbb{Z}^{n \times n}$ is unimodular then the inverse matrix U^{-1} is also unimodular and $det(U) = \frac{1}{det(U^{-1})}$.

Theorem 2.1.1.4: *Let $A, B \in \mathbb{R}^{m \times n}$ be two bases of a lattice. Then $L(A) = L(B)$ if and only if there exists an unimodular matrix $U \in \mathbb{Z}^{n \times n}$ such that $A = BU$.*

Proof.
"$\Rightarrow$" Assume $A = \{a_1, ..., a_n\}$ and $B = \{b_1, ..., b_n\}$ generate the same lattice then for each b_i of B, $b_i \in L(A)$. That means each b_i can be expressed as a linear combination of a_i's, i.e. $B = AU$ for some integer matrix U. Similarly $A = BV$ for some integer matrix V. We obtain $B = BVU$, which is equivalent to $B(I - VU) = 0$. The linear independency of the column vectors of B gives that it must be $I - VU = 0$, i.e. $VU = I$. We get $det(V) \cdot det(U) = det(VU) = det(I) = 1$ and since the matrices V and U have integer entries $det(V)$, $det(U) \in \mathbb{Z}$ and $det(V) = det(U) = \pm 1$, which implies that U and V are unimodular.

"$\Leftarrow$" Assume $A = BU$, for an unimodular matrix $U \in \mathbb{Z}^{n \times n}$ so we have $L(A) \subseteq L(B)$ and $B = AU^{-1}$. Since the matrix U is unimodular, the inverse matrix U^{-1} is also unimodular with $det(U^{-1}) = \frac{1}{det(U)} = \pm 1$. So $L(B) \subseteq L(A)$, which yields to $L(A) = L(B)$. $\square$

For $n \geq 2$ there exist infinitely many unimodular matrices U. So the bases of a lattice are not uniquely determined and we have infinitely many bases.

Definition 2.1.1.5 (Determinant of Lattice): *The determinant of a lattice L is defined as the volume of the fundamental parallelepiped*

$$det(L) = vol(P(B)) = \sqrt{det(B^\top B)}.$$

For a full-rank lattice we have $det(L) = vol(P(B)) = |det(B)|$.

Theorem 2.1.1.5: *Let A, B be bases of the same lattice, i.e. $L(A) = L(B)$. Then the determinant is well-defined $det(L(A)) = det(L(B))$, all the bases of a given lattice have the same determinant.*

Proof. Assume $L(A) = L(B)$ then by Theorem 2.1.1.4, $A = BU$ in which U is an unimodular matrix. Using this we get

$$det(A^\top A) = det((BU)^\top BU)$$
$$= det(U^\top B^\top BU)$$
$$= det(B^\top B)det(U)^2.$$

Since $det(U) = \pm 1$ we have $det(L(A)) = \sqrt{det(A^\top A)} = \sqrt{det(B^\top B)} = det(L(B))$.

$\square$

For lattice-based cryptosystems we will need vectors of a basis as orthogonal as possible. The orthogonality-defect provides a measure to detect the distance between a given basis and an orthogonal.

Definition 2.1.1.6 (Orthogonality-defect): *Given a basis $B = \{b_1, ..., b_n\}$ of a lattice L, the orthogonality-defect of it is defined as*

$$\text{orthog-def(B)} \;=\; \frac{\prod\limits_{i=1}^{n} \|b_i\|}{det(L)} \quad \text{where } \|\cdot\| \text{ denotes the euclidean norm.}$$

Theorem 2.1.1.6: *The orthogonality-defect is equal to 1 if and only if the vectors of the bases are othogonal to one another. Otherwise, the orthogonality-defect is greater than 1.*

Proof. To show that $\frac{\prod\limits_{i=1}^{n} \|b_i\|}{det(L)} \geq 1$ we have to show $det(L) \leq \prod\limits_{i=1}^{n} \|b_i\|$. Using Gram-Schmidt-Orthogonalization (GSO) we obtain a set of orthogonal vectors $b_1^*, ..., b_n^*$ where $b_1^* = b_1$ and $b_i^* = b_i - \sum\limits_{i=1}^{n-1} \frac{\langle b_i^*, b_i \rangle}{\langle b_i^*, b_i^* \rangle} b_i^*$ such that $span(b_1, ..., b_n) = span(b_1^*, ..., b_n^*)$. We can rewrite each b_i^* as $b_i^* = b_i - \sum\limits_{j=1}^{i-1} \frac{\langle b_j^*, b_j \rangle}{\langle b_j^*, b_j^* \rangle} b_j^*$ so we get $b_i = \sum\limits_{j=1}^{i-1} \frac{\langle b_j^*, b_j \rangle}{\langle b_j^*, b_j^* \rangle} b_j^* + b_i^*$ which can be expressed as $b_i = \sum\limits_{j=1}^{i-1} \mu_j b_j^* + b_i^*$ for some real number μ_j.

This definition and the orthogonality of the GSO basis result in $\|b_i\| = \sum_{j=1}^{i-1} \mu_j \|b_j^*\| + \|b_i^*\|$, with $\|b_i\| \geq \|b_i^*\|$. By Definition 2.1.1.5 we have $det(L) = vol(P(B))$ and knowing that $vol(P(B)) = \prod_{i=1}^{n} \|b_i^*\|$ yields to $det(L) = vol(P(B)) = \prod_{i=1}^{n} \|b_i^*\| \leq \prod_{i=1}^{n} \|b_i\|$.

$\square$

Example 2: *Let* $B_1 = \begin{bmatrix} 3 & 1 \\ 1 & 3 \end{bmatrix}$ *and* $B_2 = \begin{bmatrix} 1 & 0 \\ 0 & 1 \end{bmatrix}$ *be bases of the lattices* L_1 *and* L_2. *Clearly the second matrix is an orthogonal matrix. We have*

$$det(L_1) = 8,$$

$$det(L_2) = 1,$$

$$orthog\text{-}def(B_1) = \frac{\prod_{i=1}^{n} \|b_i\|}{det(L)} = \frac{\left\| \begin{pmatrix} 3 \\ 1 \end{pmatrix} \right\| \cdot \left\| \begin{pmatrix} 1 \\ 3 \end{pmatrix} \right\|}{8} = 1,25,$$

$$orthog\text{-}def(B_2) = \frac{\prod_{i=1}^{n} \|b_i\|}{det(L)} = \frac{\left\| \begin{pmatrix} 1 \\ 0 \end{pmatrix} \right\| \cdot \left\| \begin{pmatrix} 0 \\ 1 \end{pmatrix} \right\|}{1} = 1.$$

So minimizing the orthogonality-defect yields to find a basis with almost orthogonal vectors.

Definition 2.1.1.7 (Dual Lattice): *The dual lattice of a lattice* $L \subset \mathbb{R}^m$ *with basis* $B = \{b_1, ..., b_n\}$ *is defined as*

$$L^* = \{y \in span(L) \mid \langle x, y \rangle \in \mathbb{Z}, \forall x \in L\}$$

$$= \{y \in span(L) \mid \langle b_i, y \rangle \in \mathbb{Z}\}.$$

The basis of the dual lattice L^* is $B(B^\top B)^{-1}$. For a full rank lattice the basis of the dual lattice is given by $(B^\top)^{-1}$.

Theorem 2.1.1.7: *Let L be a lattice and $L^\star$ the corresponding dual lattice. Then*

$$det(L^\star) = \frac{1}{det(L)}.$$

Proof. Since $L^\star$ has basis $B(B^\top B)^{-1}$ we get

$$det(L^\star) = \sqrt{det((L^\star)^\top L^\star)} = \sqrt{det((B(B^\top B)^{-1})^\top B(B^\top B)^{-1})}$$
$$= \sqrt{det((B^\top B)^{-1}(B^\top B)(B^\top B)^{-1})} = \sqrt{det((B^\top B)^{-1})}$$
$$= \frac{1}{\sqrt{det(B^\top B)}} = \frac{1}{det(L)}.$$

For a full rank lattice L we have $det(L) = |det(B)|$ and the corresponding dual lattice $L^\star$ has basis $(B^\top)^{-1}$ so we have

$$det(L^\star) = |det((B^\top)^{-1})| = \left| \frac{1}{det(B^\top)} \right|$$
$$= \frac{1}{|det(B)|} = \frac{1}{det(L)}.$$

$\square$

Example 3: *Consider the lattice spanned by $B = \begin{bmatrix} 2 & 0 \\ 0 & 2 \end{bmatrix}$. The lattice generated by is $2 \cdot \mathbb{Z}^2$ so any $y \in L^\star$ must be such that $\left\langle y, \begin{pmatrix} 2 \\ 0 \end{pmatrix} \right\rangle$ and $\left\langle y, \begin{pmatrix} 0 \\ 2 \end{pmatrix} \right\rangle \in \mathbb{Z}$. The corresponding dual lattice $L^\star$ is $\frac{1}{2}\mathbb{Z}^2$.*

Definition 2.1.1.8 (Dual-orthogonality-defect): *Let B be a non-singular $n \times n$ matrix. Then the dual orthogonality-defect of B is defined as*

$$orthog\text{-}def^\star(B) = \frac{\prod\limits_{i=1}^{n} \|\hat{b}_i\|}{det(B^{-1})} = det(B) \prod\limits_{i=1}^{n} \|\hat{b}_i\|$$

where $\|\hat{b}_i\|$ is the euclidean norm of i-th row in B^{-1}.

After mentioning the most important properties of lattices, we focus on some problems that arise in connection with lattices.

2.1.2 Lattice Problems

In this chapter, we will describe some lattice-problems and analyze the complexity of each problem. The security of lattice-based cryptosystems depends on solving these problems.

The Shortest Vector Problem (SVP)

Let $L \subset \mathbb{R}^m$ be a lattice which contains at least one non-zero vector. A shortest vector is a lattice vector $v \neq 0$ such that $\|v\| \leq \|w\|$ for any vector $w \in L \backslash \{0\}$, $v \neq w$.

Note that the euclidean norm is referred to as the l_2-norm. The l_∞-norm is $\|x\|_\infty = \max\{x_i \mid x = (x_1, ..., x_n)\}$.

Definition 2.1.2.1 (First Successive-Minimum): *Let $L \subset \mathbb{R}^m$ be a lattice of rank n, the first successive minimum $\lambda_1(L)$ w.r.t the l_2-norm is defined as the length of one of the shortest vectors.*

Theorem 2.1.2.1: *Let $B = \{b_1, ..., b_n\}$ be a basis of a lattice $L \subset \mathbb{R}^m$ and $B^* = \{b_1^*, ..., b_n^*\}$ be its GSO. Then*

$$\lambda_1(L) \geq \min\{ \|b_i^*\|, ..., \|b_n^*\| \} > 0.$$

Proof. Let $x \in \mathbb{Z}^n$ be any non-zero vector. Then a lattice vector v can be expressed as $v = Bx$. We have to show that $v = Bx \in L(B)$ has length at least $\min\{ \|b_i^*\|, ..., \|b_n^*\| \}$. Let $j = \{1, ..., n\}$ be the largest index such that $x_j \neq 0$. We can express v as $\sum_{i=1}^{j} x_i b_i$ and since the inner product is linear we obtain

$$\left| \langle Bx, b_j^* \rangle \right| = \left| \left\langle \sum_{i=1}^{j} x_i b_i, b_j^* \right\rangle \right| = \left| \sum_{i=1}^{j} x_i \langle b_i, b_j^* \rangle \right|.$$

Because of the orthogonality of the b_i's, we have $\langle b_i, b_j^* \rangle = 0$ for $i < j$ and since $\langle b_j, b_j^* \rangle = \langle b_j^*, b_j^* \rangle$ it follows that

$$\left| \sum_{i=1}^{j} x_i \langle b_i, b_j^* \rangle \right| = |x_j| \langle b_j^*, b_j^* \rangle = |x_j| \cdot \|b_j^*\|^2.$$

The Cauchy-Schwarz inequality gives $\left|\langle Bx, b_j^* \rangle\right| \leq \|Bx\| \cdot \|b_j^*\|$, so $\frac{\left|\langle Bx, b_j^* \rangle\right|}{\|b_j^*\|} \leq \|Bx\|$. Putting all these together yields

$$\|Bx\| \geq \frac{\left|\langle Bx, b_j^* \rangle\right|}{\|b_j^*\|} = |x_j| \cdot \|b_j^*\| \geq \|b_j^*\| \geq \min_{i=1}^{n}\{ \|b_i^*\| \}.$$

Any lattice vector y has length at least $\min\{ \|b_i^*\|, ..., \|b_n^*\| \}$.

$\square$

Example 4: *Using the lattice spanned by* $B = \begin{bmatrix} 3 & 1 \\ 1 & 3 \end{bmatrix}$, *the GSO leads to the vectors* $b_1^* = \begin{pmatrix} 3 \\ 1 \end{pmatrix}$ *and* $b_2^* = \frac{1}{5}\begin{pmatrix} -4 \\ 12 \end{pmatrix}$ *with* $\|b_1^*\| = \sqrt{10}$ *and* $\|b_2^*\| = \sqrt{\frac{160}{25}}$. *Here a shortest vector is* $\begin{pmatrix} -2 \\ 2 \end{pmatrix}$ *(see Fig. 2.4) with* $\lambda_1(L) = \sqrt{8} > \sqrt{\frac{160}{25}} = \min\{ \|b_i^*\|, ..., \|b_n^*\| \}$.

Using the Minkowski inequality we also have an upper bound on the lenght of a shortest vector

$$\lambda_1(L) \leq \sqrt{\delta_n} \cdot det(L)^{\frac{1}{n}}$$

where δ_n is the Hermite's Constant. The Hermite's Constant is known for $n = 1, ..., 8$. Since each lattice is discrete there always exists a shortest vector.

The Closest Vector Problem (CVP)

Let $L(B) \subset \mathbb{R}^m$ be a lattice and $t \in \mathbb{R}^m$ be a target vector which is not necessarily in L. Find the vector $v \in L$ closest to t, i.e. such that $\|t - v\|$ is minimal.

Example 5: *In Fig. 2.5 the closest vector to the given target vector t is* $v = \begin{pmatrix} 4 \\ 4 \end{pmatrix}$.

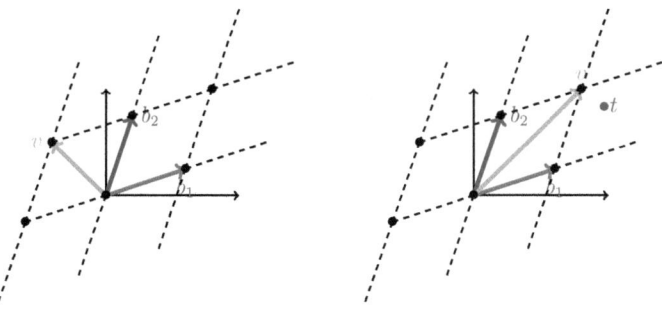

Figure 2.4: v is a shortest
vector

Figure 2.5: v is the closest
vector to t

The Smallest Basis Problem (SBP)

Let B be a basis of a lattice. The goal is to find the smallest basis B' w.r.t the small orthogonality-defect.

Example 6: *Fig. 2.6 shows the smallest basis* $B_2 = \begin{bmatrix} -2 & 3 \\ 2 & 1 \end{bmatrix}$ *for the given lattice.*
We have orthog-def(B_2) $= 1, 12 < 1, 25 = $ orthog-def(B_1) .

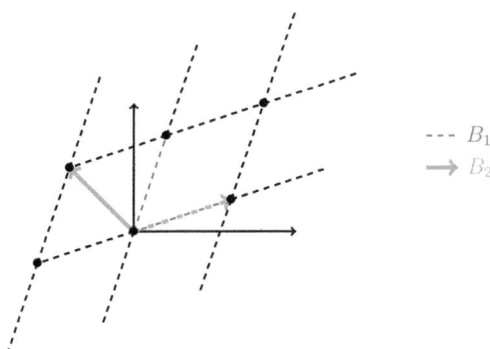

Figure 2.6: Smallest basis of given lattice

Complexity

The complexities of the lattice problems SVP and CVP were first analyzed by *Van Emde Boas* in 1981 [3]. He proved that SVP is NP-hard in the l_∞ norm and CVP is NP-hard for any l_p norm. Subsequently in [4], *Ajtai* showed that the SVP is NP-hard for the l_2 norm under randomized reductions. The SBP is also known to be NP-hard.

It is not important to solve a lattice problem exactly, but rather to find an approximation to the factor $\gamma(n) > 1$.

After many various proofs *Schnorr* [5] proved that CVP in $\mathbb{R}^n$ can be approximated to the factor $(1 + \varepsilon)^n$ for any $\varepsilon > 0$. *Micciancio* showed that approximating the SVP to within a factor of $\sqrt{2}$ is NP-hard [6].

Best polynomial-time algorithm for approximating these problems are based on the LLL-algorithm [7] and its variant. The LLL-algorithm can achieve an approximation ratio of $2^{O(n^2)}$ in the worst-case for SBP in $\mathbb{R}^n$.

Note that CVP is at least as hard as SVP, which is shown by *Goldreich* and *Micciancio* in [8]. So SVP can be solved easily by using the solution for CVP.

2.1.3 Lattice-Reduction

We know that for a lattice there exist infinitely many bases and that the orthogonality-defect is a measure to find bases with almost orthogonal vectors.

In 1982 *A. Lenstra*, *H. Lenstra* and *L. Lovász* published the LLL-algorithm. The first use of this algorithm was factorizing polynomials with rational coefficients. It turned out that this algorithm can also be used to perform lattice basis reduction in polynomial-time, which was very important at the time.

The algorithm approximates the successive minima and generates a basis consisting of short and almost orthogonal vectors. In other words the LLL-algorithm finds a LLL-reduced basis.

Definition 2.1.3.1 (LLL-reduced basis): *A basis* $B = \{b_1, ..., b_n\} \subseteq \mathbb{R}^m$ *of a lattice is LLL-reduced with parameter* δ $(\frac{1}{4} < \delta \leq 1)$ *if*

I $|\mu_{i,j}| \leq \frac{1}{2}$, *for* $1 \leq j < i < n$ *where* $\mu_{i,j} = \frac{<b_i, b_j^*>}{<b_j^*, b_j^*>}$

II   *for any vectors* b_i, b_{i+1}, $\forall 1 \leq i < n$, *we have* $\delta \|\pi_i(b_i)\|^2 \leq \|\pi_i(b_{i+1})\|^2$
where π_i *is the projection onto* $\sum\limits_{j \geq i} \mathbb{R}b_j^*$.

The first point leads to a basis with vectors as close as possible to the vectors of GSO, which is also called size reduction. The size-reduction is done if $|\mu_{i,j}| > \frac{1}{2}$ by replacing the i-th basis vector as $b_i = b_i - \lfloor \mu_{i,j} \rfloor b_j$. The LLL-algorithm uses the size-reduction algorithm and afterwards checks the second property and generates a basis where the vectors are small enough to approximate the shortest vector. The factor depends exponentially on the rank of the lattice.

Algorithm 1 LLL-Algorithm

Input: a basis $b_1, ..., b_n \in \mathbb{R}^m$
       a parameter δ
Output: δ-LLL-reduced basis

  compute $b_1^*, ..., b_n^*$
  $B_i = \|b_i^*\|^2$
  if $\mu_{k,k-1} > \frac{1}{2}$ **then**
    $sizereduce(b_k)$
  end if
  if $\delta \|b_{k-1}^*\|^2 > \|b_k^*\|^2 + \mu_{k,k-1}^2 \|b_{k-1}^*\|$ **then**
    $swap(b_k, b_{k-1})$
    $k = max\{k - 1, 2\}$
  else
    $k = k + 1$
  end if

Example 7: *To demonstrate how this algorithm works consider this example:*

Given lattice spanned by $B = \begin{bmatrix} 2 & -4 \\ 4 & -2 \end{bmatrix}$ *and parameter* $\delta = \frac{3}{4}$. *We have* $|\mu_{2,1}| = \frac{4}{5} > \frac{1}{2}$ *thus size reducing* b_2 *leads to the vector* $b_2 = (-2, 2)^\top$. *Using GSO we get* $b_1^* = \begin{pmatrix} 2 \\ 4 \end{pmatrix}, b_2^* = \begin{pmatrix} 12/15 \\ 6/5 \end{pmatrix}$, *with* $\|b_1^*\|^2 = 20$ *and* $\|b_2^*\|^2 = \frac{36}{5}$. *Since* $\delta\|b_1^*\|^2 = 15 > \|b_2^*\| + \mu_{2,1}^2\|b_1^*\|^2 = 8$, *we have to swap* b_1 *and* b_2: $\begin{bmatrix} -2 & 2 \\ 2 & 4 \end{bmatrix}$.

Applying GSO again leads to $b_1^* = \begin{pmatrix} -2 \\ 2 \end{pmatrix}, b_2^* = \begin{pmatrix} 3 \\ 3 \end{pmatrix}$, *with* $\|b_1^*\|^2 = 8$ *and* $\|b_2^*\|^2 = 18$. *Checking the property of size-reducing again gives* $\mu_{2,1} = \frac{1}{2}$, *which means the first condition is satisfied. Since* $\delta\|b_1^*\|^2 = 6 < \|b_2^*\| + \mu_{2,1}^2\|b_1^*\|^2 = 20$ *the algorithm terminates.*

The LLL-reduced basis is $\begin{bmatrix} -2 & 2 \\ 2 & 4 \end{bmatrix}$.

Using this algorithm one can find a relatively short vector in polynomial-time. A short vector in this examle is $v = (-2, 2)^\top$ which is also the first row of the LLL-reduced basis (see Fig.2.7).

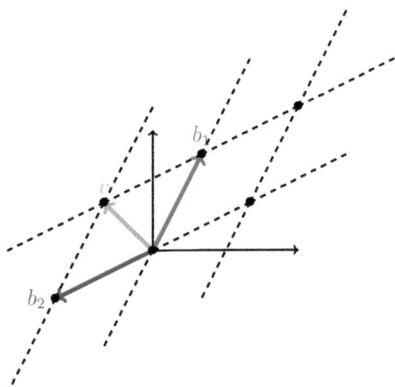

Figure 2.7: A shortest vector v of given lattice

2.2 Asymmetric Cryptosystems and Digital Signatures

2.2.1 Asymmetric Cryptography

Cryptographies are used to send messages securely over a channel. This is done by encrypting the plaintext such that decrypting the ciphertext is only allowed by the legitimate recipient. In former times, the symmetric-key encryption was used, where the key for encryption and decryption is the same. The sender of a message and the recipient had to share a key and keep it secret. But sharing the secret-key is a security problem. Another point is that for n parties $\frac{n(n-1)}{2}$ keys are needed, which creates a key management problem.

In 1976 *Diffie* and *Helman* [9] published the asymmetric cryptography (public-key), where one can send also a message securely but without the need of sharing a key with anyone. In asymmetric cryptography we have to create a key-pair where one key is used for the encryption and the other one for decryption.
There is a public key to identify the trapdoor which can only be unlocked by the legitimate recipient by using a private key, which is kept secret.
The public key of the recipient can be known and can be used by any sender to encrypt a message to him. It must be computationally infeasible to derive the private key from the known public key. This is based on a one-way trapdoor function $f(x)$ which is easy to compute for given x, but it is hard to compute x from $f(x)$ in polynomial-time.

An example for the asymmetric cryptography is the RSA:

Key generation: pick 2 prime numbers p, q and compute $n = pq$
    public key: (e, n) where e is an integer such that $gcd(e, p - 1) = 1$ and
    $gcd(e, q - 1) = 1$
    private key: (d, n) where d is an integer such that $d \cdot e = 1 \bmod lcm(p - 1, q - 1)$

Encryption: $c = m^e \bmod n$

Decryption: $m = c^d \bmod n$

Clearly, if the modules $n = p \cdot q$ can be factored, the security of RSA fails. If one can retrieve the primes p and q he can retrieve the secret exponent by using p, q to invert the public key exponent.

2.2.2 Digital Signatures

The public-key cryptography can also be used to authenticate the sender of a message such that the recipient can be sure that the message was not change on the way and that the message is really sent by the alleged sender.
The private key is known only to the signer while the public key is typically stored in a certificate, which is signed securely by the Certificate Authority. The signer can use his private key to create the signature and the public key is used to verify the signature.

First, the message m must be hashed then the sender can sign using his private key. So he get a digital signature which is attached to the message and sent with it. The recipient gets both, the message and the digital signature, and can verify the signature using the public key from certificate.
The receiver must hash the message as well and then compare the resulting $h(m')$ and the hashed message $h(m)$. If the messages are the same, then the signature is valid.

3 Lattice-based Cryptosystems

This chapter will give a description of the lattice-based cryptosystems, which is the actual goal of the previously introduced principles and problems.

The lattice-based cryptosystems are public-key systems and work with full rank lattices. As already mentioned, the security of some cryptosystems is based on solving some lattice problems.

The system we will explain in detail is the GGH-cryptosystem. It also allows to provide a digital signature which plays an important role in secure transactions. First, the construction and the security of the GGH-cryptosystem will be described and analyzed. Finally, it will be compared to other cryptosystems.

3.1 GGH-Cryptosystem

Goldreich, Goldwasser and *Halevi* presented in 1997 the GGH-cryptosystem [10], which is based on a trapdoor one-way function depending on the difficulty of CVP. This trapdoor function was used to create a public-key cryptosystem and digital signatures.

The idea is to use a basis of a lattice as public key and a reduced basis of the same lattice as private key. Given any basis for a lattice, it is easy to create a vector close to a lattice vector, but hard to derive the original vector from this generated lattice vector.

3.1.1 Construction

Key Generation

We need a basis B and a reduced basis R of the same lattice such that R has a low dual-orthogonality-defect and B has a high dual-othogonality-defect.
We first create the reduced basis and then generate the lattice.

There exist two ways to choose R. The simplest way is to choose the entries of the matrix uniformly at random from $\{-l, ..., l\}^{n \times n}$ for some integer bound l (e.g. $l = 4$). The other way is to generate an almost rectangular lattice, starting from $k \cdot l \in \mathbb{R}^n$ and adding noise to each of the vectors. We generate $R = R' + k \cdot l$ where R' is an uniformly distributed matrix in $\{-l, ..., l\}^{n \times n}$. The best k is suggested to be about $\sqrt{n} \cdot l$.

After creating the private basis R we have to generate the public key B from R. By Theorem 2.1.1.4 we know that two bases A and B can generate the same lattice if and only if $A = BU$ for a unimodular matrix U. So we multiply the private basis R by unimodular matrices $U_i = L_i \cdot T_i$, where L_i and T_i are upper and lower triangular matrices with diagonal entries in $[-1, 1]$. This leads to vectors which are larger than the entries of the private basis. At least 4 unimodular matrices are required to prevent recovering the private basis R.

To get a public basis B, we can also transform R by taking one vector of the basis and adding to it a random integer linear combination of the other vectors.

Encryption/Decryption

To encrypt a message we use the public basis B to create a random vector $v \in L$. Let $B \in \mathbb{Z}^{n \times n}$ be the public basis and $p \in \mathbb{Z}^n$ the plaintext.

Given a positive real number σ, a randomly choosen vector e from $\{-\sigma, \sigma\}^n$ and a lattice-vector v we can encrypt the plaintext as

$$c = v + e$$

such that p can be uniquely determined by $v = B \cdot p$, so we have $c = Bp + e$.

The decryption is based on the assumption that v is the closest vector to c. We use the private basis R, which is the "good basis", to find the closest vector to c.

Given the private basis R and the ciphertext c we have to approximate the CVP using the Round-off method of Babai [11].

Due to this method we first compute $R^{-1}c$ then we round the entries and multiply with R. Then we get $v = R \lfloor R^{-1}c \rceil$, which is the closest vector to c.

The Round-off method is based on the assumption that $\lfloor R^{-1}e \rceil = 0$.

With this assumption we get

$$
\begin{aligned}
R\lfloor R^{-1}c\rceil &= R\left\lfloor R^{-1}Bp + R^{-1}e\right\rceil \\
&= R\left\lfloor R^{-1}RUp + R^{-1}e\right\rceil \\
&= R\left\lfloor Up + R^{-1}e\right\rceil \\
&= R\lfloor Up\rceil + R\underbrace{\left\lfloor R^{-1}e\right\rceil}_{=0} = Bp.
\end{aligned}
$$

Since U is an integer matrix, rounding the entries leads to Up. This means the ciphertext c can be decrypted as

$$
p = B^{-1}R\lfloor R^{-1}c\rceil
$$

such that it can be completed using B, R and c.

In case $\lfloor R^{-1}e\rceil \neq 0$ we get an inversion error. The probability of an inversion error depends on the size of the matrix R^{-1}.

Theorem 3.1.1.1: *Let R be the private basis and R^{-1} be bounded by $\frac{\gamma}{\sqrt{n}}$. Then*

$$
Pr\left[inv.\ error\right] \leq 2n \cdot exp\left(-\frac{1}{8\sigma^2\gamma^2}\right).
$$

Proof. For the proof we will need the Hoeffdin inequality:
Let $x_1, ..., x_n$ be independent random variables and bounded, $a_i \leq x_i - E[x_i] \leq b_i$. Then for any positive constant c it yields

$$
Pr\left[\sum_{i=1}^{n}(x_i - E[x_i]) \geq c\right] \leq exp\left(-\frac{2c^2}{\sum\limits_{i=1}^{n}(b_i - a_i)^2}\right).
$$

Let $d = R^{-1}e$ and denote its i-th entry by δ_i. Denote the i-th entry in e by ε_i and i-th row in R^{-1} by r_i. We have $\delta_i = r_i \circ e = \sum\limits_{j=1}^{n} \beta_{ij}\varepsilon_j$ where β_{ij} is the i, j-th element in R^{-1}. Set $Pr\left[|\delta_i| \geq \frac{1}{2}\right]$ for some i, $1 \leq i \leq n$, the inversion error occurs if $|\delta_i| > \frac{1}{2}$. We have $Pr\left[inv.\ error\right] \leq n \cdot Pr\left[|\delta_i| \geq \frac{1}{2}\right]$. The entries of ε_j are uniformly distributed in $\{\pm\sigma\}$ so $E[\varepsilon_j] = 0$ and thus $E[\beta_{ij}\varepsilon_j] = 0$.

According to the condition we have $|\beta_{ij}| \leq \frac{\gamma}{\sqrt{n}}$ $\forall j$ such that $-\sigma\frac{\gamma}{\sqrt{n}} \leq \beta_{ij}\varepsilon_j \leq \sigma\frac{\gamma}{\sqrt{n}}$. Using the Hoeffding inequality it follows

$$Pr\left[|\delta_i| > \frac{1}{2}\right] = 2 \cdot Pr\left[\delta_i > \frac{1}{2}\right] = 2 \cdot Pr\left[\sum_{i=1}^{n}\beta_{ij}\varepsilon_j > \frac{1}{2}\right] \leq 2exp\left(-\frac{1}{8\sigma^2\gamma^2}\right).$$

$\square$

This leads to choose the σ less than $\left(\gamma\sqrt{8ln(\frac{2n}{\varepsilon})}\right)^{-1}$. A given example in their paper is that for $n = 120$, $\varepsilon = 10^{-5}$ and $l = 4$ yields $\gamma = \frac{1}{30}$ and thus $\sigma \leq 2, 5$.

Example 8: *For simplicity, generate the public basis B using only one unimodular matrix. Consider $\sigma = 1$, R generated randomly from $\{-4, ..., 4\}$*

$$R = \begin{bmatrix} 3 & -4 & -1 \\ 3 & 1 & 2 \\ 1 & -3 & 3 \end{bmatrix} \quad U = \begin{bmatrix} 1 & -1 & 1 \\ 0 & 1 & -1 \\ 1 & 1 & 0 \end{bmatrix} \quad B = R \cdot U = \begin{bmatrix} 2 & -8 & 7 \\ 5 & 0 & 2 \\ 4 & -1 & 4 \end{bmatrix}$$

Let $p = (2, 7, -3)^{\top}$ be the plaintext we want to encrypt and $e = (-1, 1, 1)^{\top}$ be a randomly choosen vector from $\{-\sigma, \sigma\}$. Then we get

$$c = Bp + e = \begin{pmatrix} -73 \\ 4 \\ -11 \end{pmatrix} + \begin{pmatrix} -1 \\ 1 \\ 1 \end{pmatrix} = \begin{pmatrix} -74 \\ 5 \\ -10 \end{pmatrix}$$

The ciphertext is $c = (-74, 5, -10)^{\top}$ and the closest vector to it is $v = (-73, 4, -11)^{\top}$.

Now we should be able to reconstruct our plaintext using B, R and c. First we solve the CVP using the Rounding method:

$$\lfloor R^{-1}c \rceil = \left\lfloor \begin{bmatrix} 9/65 & 3/13 & -7/65 \\ -7/65 & 2/13 & -9/65 \\ -2/13 & 1/13 & 3/13 \end{bmatrix} \begin{pmatrix} -74 \\ 5 \\ -10 \end{pmatrix} \right\rceil = \begin{pmatrix} -8 \\ 10 \\ 9 \end{pmatrix}$$

and applying R:

$$R\lfloor R^{-1}c\rceil = \begin{bmatrix} 3 & -4 & -1 \\ 3 & 1 & 2 \\ 1 & -3 & 3 \end{bmatrix} \begin{pmatrix} -8 \\ 10 \\ 9 \end{pmatrix} = \begin{pmatrix} -73 \\ 4 \\ -11 \end{pmatrix} = v'$$

Using this vector we decrypt c as

$$p' = B^{-1}v' = \begin{bmatrix} 2/65 & 5/13 & -16/65 \\ -12/65 & -4/13 & 31/65 \\ -1/13 & -6/13 & 8/13 \end{bmatrix} \begin{pmatrix} -73 \\ 4 \\ -11 \end{pmatrix} = \begin{pmatrix} 2 \\ 7 \\ -3 \end{pmatrix}$$

which is equal to p.
Check whether p′ is decrypted without an inversion error:

$$\lfloor R^{-1}e\rceil = (0,0,0).$$

Signing/Verifying

Signing and Verifying is similar to the Encryption scheme.

Let B be a basis and R be a reduced basis of the same full rank lattice in $\mathbb{Z}^n$. To prove his identity, the signer has to solve the CVP by using the basis R, which is known only to him.

For the plaintext p the signer must find a lattice vector v that is close to the vector p by using Babai's algorithm. Then the vector $v = R\lfloor R^{-1}p\rceil$ is the signature.

The signature is verified by showing that v is a lattice vector and is the closest vector to the message vector by using the basis B, i.e. we have to show that $\exists x \in \mathbb{Z}^n$, s.t. $v = Bx$ and that $\|p - v\|$ is sufficiently small.

3.1.2 Attacks

In the paper of *Goldreich, Goldwasser* and *Halevi*, attacks on the encryption scheme were presented as well. They use a public basis B' for the attacks, which is already reduced via LLL-algorithm.

The Round-off attack, Nearest-plane attack and the Embedding attack were analyzed in various dimensions. The results show that all attacks become infeasible in dimension above 150.

Round-off attack

In this attack we assume that one try to decrypt the message without knowing the private basis R.

Given the ciphertext $c = Bp + e$ we compute

$$B^{-1}c = p + B^{-1}e$$

So the attacker must find the vector $d = B^{-1}e$ to get the correct plaintext. For the vector d we can do a brute-force search. The authors have shown that the search space of the vector d is exponential in the dimension of the lattice and that the work-load of this attack grows by factor of about 8000 per dimension. Already in dimension 100 this attack is worse than brute-force search for the vector e.

Nearest-plane attack

Another possible way to approximate the CVP is the use of the Nearest-plane algorithm, which is also presented by *Babai*.

Let $B = \{b_1, ..., b_n\}$ be a LLL-reduced basis, the algorithm finds a lattice vector v close to the given vector $c \in \mathbb{R}^n$ by projecting c onto the hyperplane $H + u$, $u \in L$ where $H = span\{b_1, ..., b_n\}$ and is the closest to c.

So we get a new vector $c' \in H + u$ such that $c - c' \in H^\perp$ and a new basis $B' = \{b_1, ..., b_{n-1}\}$. The algorithm computes recursively a vector c'' close to $c' - u$. If $c \notin L$ then $c'' \notin L$. The solution of CVP is $v = c'' + u$.

In this attack we can also try the second closest hyperplane H, it must not be the closest one.

The work-load of this attack is much lower than the Round-off attack but grows exponentially with dimension of the lattice and becomes infeasible in dimension 140-150.

Embedding attack

Embedding Attack is the method to embed the vector c and the n vectors of the known basis B in a $(n + 1)$-dimension lattice $L(B')$.

We get $p = B^{-1}(c - e)$ with $B' = \{(c, 1), (b_1, 0), ..., (b_n, 0)\}$. Then we apply the LLL-algorithm to find the shortest non-zero vector in $L(B')$.

This works up to dimensions about 110-120.

Example 9: *Given* $B = \begin{bmatrix} 2 & -8 & 7 \\ 5 & 0 & 2 \\ 4 & -1 & 4 \end{bmatrix}$ *and the ciphertext* $c = (-74, 5, -10)^\top$, *we*

embed the vector c as $B' = \begin{bmatrix} -74 & 2 & -8 & 7 \\ 5 & 5 & 0 & 2 \\ -10 & 4 & -1 & 4 \\ 1 & 0 & 0 & 0 \end{bmatrix}$ *and get* $B' = \begin{bmatrix} 1 & 0 & 0 & 0 \\ 0 & -1 & -1 & 2 \\ 0 & 5 & 0 & 2 \\ 0 & 0 & -5 & -1 \end{bmatrix}$ *after*

applying the LLL-algorithm.

The first row $(1, 0, 0)$ *is the difference between c and the closest vector, so we have*

$$v' = \begin{pmatrix} -74 \\ 5 \\ -10 \end{pmatrix} - \begin{pmatrix} 1 \\ 0 \\ 0 \end{pmatrix} = \begin{pmatrix} -75 \\ 5 \\ -10 \end{pmatrix}.$$

Using this vector we get

$$p' = \lfloor B^{-1} v' \rceil = \begin{pmatrix} 2 \\ 8 \\ -3 \end{pmatrix}$$

which is similar but not equal to p.

We have presented some possible attacks for the analysis of the security of the GGH-cryptosystem. Lattices with high dimensions leads to a more secure cryptosystem.

The work-loads of these attacks are compared in Table 3.1 for some dimensions. The work-load of the Embedding-attack is not known. Instead of work-load the highest value of σ for which the attack was successfull is given.

Dimension	Workload (log_2)		σ
	Round-off	Nearest-plane	Embedding
$n = 90$	72	1	0.95
$n = 110$	334	24	2,3
$n = 130$	572	63	1,8
$n = 150$	855	104	

Table 3.1: Attacks in some dimensions

3.2 Comparison to other cryptosystems

A cryptography is considered as secure if it is average-case hard, e.g. RSA requires that factoring is average-case hard under a certain distribution. But the problem of factorization can be solved in polynomial time. *Ajtai* and *Dwork* had developed for the first time a cryptosystem based on a problem for which we can show a worst-case/average-case equivalence. A cryptosystem is proven secure if a worst-case/average-case equivalence can be shown. So for the security the lattice-based cryptosystems opened a new opportunity.

Furthermore, lattice-based cryptosystems are more efficient.

To compare the lattice-based cryptosystems with each other we give a brief overview of the NTRU-cryptosystem and the cryptosystem presented by *Ajtai* and *Dwork*.

Ajtai-Dwork

After showing that the SVP is NP-hard, a cryptosystem based on such a hard problem was pointed out by *Ajtai* and *Dwork* [12].

In this cryptosystem the private key is a n-dimensional vector u which is uniformly distributed from n-dimensional unit ball. To create the public key we need $a_1, ..., a_m$ vectors from $\{x \in B \mid \langle x, u \rangle \in \mathbb{Z}\}$ then we set $\delta_i = \sum_{j=1}^{n} \delta_{ij}$ for $i = 1, ..., m$ and $v_i = a_i + \delta_i$. Let i_0 be the smallest entry in $P(v_{i+1}, ..., v_{i+n})$ where P is the parallelepiped, then $w_j = v_{i_0+j}$. The public key is $(v_1, ..., v_m)$ and i_0.

The encryption is bitwise. For $v' \in P$ we have $v' = v \bmod P \Leftrightarrow v = v' + \sum_{i=1}^{n} z_i w_i$, $z_i \in \mathbb{Z}$. To encrypt a zero bit we have to choose $b_1, ..., b_n \in \{0, 1\}$ and reduce the vector $c = \sum_{i=1}^{m} b_i v_i \bmod P$ which is finally our ciphertext.

To encrypt a one bit we choose an uniformly distributed random vector c from P.

To decrypt the ciphertext under the private key u we compute $r = \langle c, u \rangle$. If r is at most $\frac{1}{n}$ away from an integer, i.e. $dist(r, \mathbb{Z}) \leq n^{-1}$, then we decrypt c as a zero bit, otherwise as an one bit.

Note that a one bit can be decrypted incorrectly as a zero bit with probability $\frac{2}{n}$.

The authors showed that this cryptosystem require solving the unique shortest vector problem (USVP). If one can distinguish the encryption of a zero bit from an one bit encryption in polynomial time, then the USVP has a polynomial time solution in the worst-case.

NTRU

The NTRU cryptosystem was published in 1996 by *Silverman, Hoffstein, Pipher* and *Lieman* [13]. In this cryptosystem the encryption is similar to the encryption in GGH-cryptosystem but decryption is not based on solving the CVP.

The private key is a pair of polynomials f and g such that $f_q^{-1} \circledast f = 1 \ (mod \ q)$ and $f_p^{-1} \circledast f = 1 \ (mod \ p)$. Then the public key is computed as $h = f_q \circledast g \ (mod \ q)$.

We put the message m in polynomial form with coefficients $\{-1, 0, 1\}$ and encrypt as $c = pr \circledast h + m \ (mod \ q)$, where r is a randomly choosen polynomial. To decrypt the ciphertext c we have to compute $a = f \circledast c \ (mod \ q) = pr \circledast g + f \circledast m \ (mod \ q)$, a choosen in $[-\frac{q}{2}, \frac{q}{2}]$. Using this we get $m' = f_p^{-1} \circledast a \ (mod \ p) = f_p^{-1} \circledast pr \circledast g + m \ (mod \ p) = m \ mod \ p$, since $pr \circledast g \ (mod \ p) = 0$ and $f \circledast f_p^{-1} = 1 \ mod \ p$.

This encryption scheme can also be expressed using lattices, since the convolution product of polynomials is equal to the multiplication with a circulant matrix. The private vectors f and g are lattice-vectors and the vector (f, g) is probably a shortest vector of the lattice.

On NTRU are also several attacks possible, i.a. lattice-based attacks. A basis B using the public key can be created such that one can try to find the shortest vector in this lattice via lattice-reduction. If one can find the vector (f, g) he can decrypt the message.

A short summing-up shows the following picture:

The security of the GGH-cryptosystem is based on solving the SBP in case deriving the private basis R from the public basis B, and on solving the CVP in case deriving the vector $v = Bp$ from the ciphertext c. The security of the NTRU-cryptosystem is based on solving the SVP. Both systems work good only with high dimension lattices. The higher the dimension, the more secure are the systems.

The security of the Ajtai-Dwork cryptosystem is based on solving the USVP, which seems to be easier than the SVP. This system is not very practical and is less efficient. It was only theoretically important.

In any case lattice-based cryptography is proven secure in consideration of the worst-case hardness of the lattice-problems.

The computation time for NTRU and GGH are quadratic in the security parameter n. For RSA the computation time is $O(n^3)$.

A small disadvantage of the GGH and Ajtai-Dwork cryptosystems is the size of the keys which is $O(n^2)$ whereas in NTRU and RSA the size is only $O(n)$. The private key in Ajtai-Dwork cryptosystem is a n-dimensional vector and the binary expression of each entry is n so n^2 bits are needed to store it.

4 Conclusion and Future Work

We have introduced the basic principles of the lattice theory, which is needed to understand the lattice-based cryptography. The complexity of lattice-problems were analyzed and we have shown that these problems are very hard to solve. We also introduced an algorithm to approximate these problems in polynomial-time. Finally a public-key cryptosystem based on solving the lattice-problem CVP was presented, which can also be used as a signature scheme. The positive and negative aspects were also discussed.

The GGH-cryptosystem is secure as long as we choose the parameters correctly and is more efficient compared to other cryptosystems. But high lattice dimensions are required and the size of keys grows quadratically in the dimension, which seems to be a disadvantage. In future an improvement in the key-size would be desirable.

Bibliography

[1]   CLAUS-PETER SCHNORR: *Gittertheorie und algorithmische Geometrie, Reduktion von Gitterbasen und Polynomidealen.* Johann-Wolfgang-Goethe University, Frankfurt/Main, 1994

[2]   DANIELE MICCIANCIO: *CSE 206A: Lattice Algorithms and Applications.* University of California, San Diego, 2014

[3]   VAN EMDE BOAS: *Another NP-complete problem and the complexity of computing short vectors in a lattice.* University of Amsterdam, 1981

[4]   MIKLÓS AJTAI: *The shortest vector problem in L_2 is NP-hard for randomized reductions.* Proc. 30th Annual ACM Symposium on Theory of Computing, 1998

[5]   CLAUS-PETER SCHNORR: *A hierarchy of polynomial time lattice basis reduction algorithms.* Theoretical Computer Science, vol. 53, 1987

[6]   DANIELE MICCIANCIO: *The shortest vector problem is NP-hard to approximate to within some constant.* SIAM Journal on Computing, 2001

[7]   A. LENSTRA, H. LENSTRA, L. LOVÁSZ: *Factoring polynomials with rational coefficients.* Math. Ann. 261, 1982

[8]   O. GOLDREICH, D. MICCIANCIO, S. SAFRA, P. SEIFERT: *Approximating shortest lattice vectors is not harder than approximating closest lattice vectors.* Inf. Process. Lett. 71, no. 2, 1999

[9]   W. DIFFIE, M. HELLMAN: *New directions in cryptography.* IEEE Transactions on Information Theory. 22, Nr. 6, 1976

[10] O. GOLDREICH, S. GOLDWASSER, S. HALEVI: *Public-key cryptosystems from lattice reduction problems.* CRYPTO '97, Springer-Verlag, 1997

[11] LÁSZLÓ BABAI: *On Lovász' lattice reduction and the nearest lattice point problem.* Combinatorica, Vol. 6, 1986

[12] M. AJTAI, C. DWORK: *A public-key cryptosystem with worst-case/average-case equivalence*. Proc. 29th Annual ACM Symposium on Theory of Computing, 1997

[13] J. HOFFSTEIN, J. PIPHER, J. SILVERMAN: *NTRU: A ring-based public key cryptosystem*. Algorithmic Number Theory (ANTS III), Springer-Verlag, 1998